HOW WE KNOW WHAT WE KNOW ABOUT
GEOLOGY

Written by:
Steven D.J. Baumann

Front Cover Photo:
Upper Falls at Hungarian Falls in the Upper Peninsula of Michigan. Photo was taken by the author on May 1, 2021.

Back Cover Photo:
The rocks along Grinnell Trail in Glacier National Park, Montana. Photo was taken by the author on September 7, 2019.

Contents

Introduction..5

How did it all get started? the founders of modern geology .11

The geologic timescale: a dynamic concept 18

Continents moving around the globe: Yeah. Right?............29

The discovery of mountain building 40

Great granite debate: from water to magma 45

Ice ages: a cold concept..52

Planetary geology: a new frontier 58

Unsolved mysteries in geology..59

Glossary of scientific terms..61

About the author 67

Acknowledgements...68

References 69

Web references...72

Photos 73

Further reading 102

Epilogue..103

Introduction

All science is a process. We begin naive and develop our ideas and techniques based off of what we observe, test, and experiment upon. We start with an idea, usually beginning with something we observed. The initial observation is a response to something we see and think, "I wonder how that came to be." We come up with an idea as to how that thing "came to be". That idea is called a hypothesis. It isn't just any explanation. It is one that can be tested, and either verified or falsified. It doesn't start at a conclusion and work backwards. We do not limit our observations to our hypothesis. If we later observe something that contradicts our hypothesis, we change the hypothesis. We do not rigidly stick to it when it does not work. If you don't follow the evidence, you have left science and entered realm of biases and ideology.

In today's world, we have seen the benefits of science, from the discovery of electricity to space travel. It is understandable that some public perception of science as established and static, has come to be. By no means is science established. Science has evolved to the point where there are well accepted theories that have never been disproven through objective observation and experimentation, so we continue to build upon what we have learned until one of two things happen. Either the theory becomes more solidified through evidence, or evidence to the contrary emerges and we go back to the drawing board. The result is ideas in science do change over time. Science is more than just a process; it also has a history to it.

If you are unfamiliar with the scientific process and why science refuses to conform to ideology, here is an example of how the process works. No one currently argues that when you combine one sodium$^+$ atom with one chlorine$^-$ atom, you get salt. This isn't an assumption or guess; this is well documented. We didn't start out thinking salt is the combination of sodium and chlorine. We started with the hypothesis that atoms form molecules if you react them together. Then we started to combine things through experimentation. All kinds of atoms were combined to form molecules. We observed that when you combine one positively charged sodium atom and one negatively charged chlorine atom you get salt, not 15% of the time, not 55.5% of the time, not 75% of the time, not 99.999% of the time. You get salt 100% of the time. This observation can be used to predict other reactions before we run the experiment. We can predict that one atom with a plus one charge combined with another atom with a minus one charge will always combine to form a neutrally charged molecule. Then we run the experiments and find out that that is indeed what happens. We made a specific prediction that came true.

Prediction is especially important in validating a scientific theory. Predictions have to be awfully specific, and they usually involve at least one equation. Vague Nostradamus style statements that can be read 50 different ways have no place in the scientific process.

The history of other sciences has undergone a similar process, but we are far from knowing everything. The vast body of what we don't know far outweighs what we actually know. There are new discoveries all the time. Science evolves over time and most have their roots in more mythical constructs. Science began to deviate from its more parent philosophies through the process of methodological naturalism and the utilization of the scientific method. Instead of trying to determine some deep absolute truth based in mythology, science focused on finding some focused natural explanation for an observation or occurrence. Alchemy became chemistry, the philosophy of the natural world became physics, astrology became astronomy, naturalism became biology, and natural history became geology.

Geology is fascinating to me, not only because I am a professional geologist, but because it is a very dynamic science with a long history of both revolutionary ideas along with very poorly constructed ones. It is important to understand that we are confident the Earth is 4.56 billion years old, plus or minus about 20 million years. We didn't arrive at that figure by "guessing". Nor did we reach it consulting contradictory ancient texts or suggestions by long dead philosophers. We came to that number after decades of tests and observations. No one woke up in the late 19th and early 20th centuries and said, "The Earth is 4.56 billion years old, now let's find evidence proving evolution, plate tectonics, ice ages, etc." That isn't how science works. That's how ideology works.

In this book, I am going to give you a brief history and overview of how certain accepted theories of geology came to be, from initial observations to the development and changing of our ideas over time. Most modern geologic theories can be traced back to field observations made in the 18th and 19th centuries; some can be traced back even further to ancient Greece. We aren't going to go that far back in this book. Herein I am just dealing with the modern evolution of geology as a science.

I just want you to keep in mind as you read this, that field observations of the geology seen on the ground did, and still does, play perhaps the most important part of what we know about how our planet evolved. Modern computers can aid us in analysis of our data. However, if computers are solely used to generate a hypothesis, and that hypothesis cannot be validated without physical evidence or predictive modeling, it should be taken with extreme skepticism.

Uniformitarianism, plutonism, and Neptunism

I will be talking not only about long abandoned concepts in geology but also old ones that we still use, sort of. The concept of uniformitarianism is one such concept. When it came about, science was not as it is today. It was tied closer to philosophy. As a result, many early concepts tend to have an "-ism" attached to the end. And the concepts weren't developed in the same manner that scientific theories are today. They were more of origin stories based on intuition and loose observations to take in tow with you while on expeditions. Many concepts were tossed around like plutonism, Neptunism, uniformitarianism, etc., but that doesn't mean one school of thought is completely correct at the expense of others.

In the early days of naturalism, we had no concept of ice ages, meteor impacts, and super volcanos. We still had this notion that the universe functions through some sort of absolute truth that we can just reason our way to discovering. That isn't how you actually understand the natural world. Although you may hear terms used like uniformitarianism, it doesn't have much meaning in modern geology. What matters is what we can model and accurately predict based on our observations and experiments. Modern geology has moved away from arguing "-isms" and focus more on the scientific method. It's one of the reasons it advances. The universe is under no obligation to conform to any person's preconceived notion of it. We can only describe what is testable based on our best techniques and technology at the time. If that produces a viable origin of how something came to be, then so be it. If it doesn't, that's fine as well.

This is not a book strictly on the history of geology and a catalog of its contributors. Its main purpose is to show you how several key concepts within geology developed over time. It is not a comprehensive bibliography on every pioneer in the field. Many key players are left out, not because they are insignificant, but because I wanted to keep this book flowing and as concise as possible. It is an introductory book that will hopefully expand your basic understanding.

How did it all get started? the founders of modern geology

James Usher

In the middle 18th century, it was accepted that the biblical account of the creation of the Earth was fact. This belief wasn't based on any observed evidence. It was based on an assumption, because they lacked anything better. It was a starting point from which they were familiar. Those who claimed the earth was young, like Archbishop James Usher (1581-1656) believed that the Earth was created on October 22, 4004 B.C. He came to this conclusion by utilizing the genealogy in the bible. He never looked at a single rock, nor did he compare the anatomy of living animals, nor did he consult with others outside the church. I am not just negatively criticizing him. He used what he thought were the proper tools during his time. At the time it was the only explanation of how the earth formed. The modern scientific method had just been documented and defined, for the first time, by Sir Francis Bacon (1561-1626) of England. As a society, Europe was just coming out of the Middle Ages.

Catastrophism

The believers of a young earth would later split into two. A third group of early naturalists such as James Hutton and Sir Charles Lyell would develop geological ideas and drift away from the literal translation of the bible altogether. Some young earth proponents would later fall into the camp of "catastrophism", the belief that catastrophes (like the biblical flood) are solely responsible for shaping the earth. The other believers in a young earth would stick rigidly to a literal interpretation of the bible and become the young earth creationists of today.

James Hutton, Sir Charles Lyell, Charles Darwin, and others would fall into a third group, the "uniformitarianism" camp. They believed that the earth was shaped over a long period of time by slow processes. In the early 21st century we realize almost all the geologic processes that shaped the earth are known to take place over vast periods of time. However, catastrophes do play an important part, such as the asteroid impact at the end of the late Cretaceous that drove the dinosaurs to extinction.

Today we know the vast majority of Earth processes take place very gradually, occasionally punctuated with impacts, eruptions, and ice ages. But nothing on the scale of a global worldwide flood event has ever been observed in the geologic record. As with most processes that shape the natural world, a combination of ideas tends to explain the observations. Very rarely can a single cause or event explain all the observable data. That's why when walking through unfamiliar territory we start off small in science and build upon successful concepts eventually forming a theory composed of observations, data, fact, predictable models, and scientific laws.

James Hutton

I've been tossing some names around like Hutton and Lyell. These two individuals are often considered the founders of modern geology. James Hutton was born in Edinburgh Scotland in 1726. He was a farmer by occupation and studied chemistry and medicine. It was the farmer in him that turned him into the first true geologist.

By the 1760s he began to make some key observations. He noticed how rocks tended to exist in layers, but he was by no means the first to notice this and write it down. That was arguably Nicolas Steno, who I will come back to later.

Hutton also observed that certain fossils only occurred in certain geologic layers or strata. He saw how certain rock types cut across other existing strata and that some layers were tilted from a horizontal position, mainly the vertical gray shale over the flat red sandstone at Siccar Point. He concluded from these observations that the processes that shaped the rocks must have taken vast amounts of time. At this point uniformitarianism was born, but untested. His greatest contribution was his observation that what we call today, sedimentary rocks, must have been deposited from the decay (erosion) of existing rocks. He called this the "great geologic cycle". We call this process today the "rock cycle". Although Hutton didn't understand the complex processes of how one rock becomes another, he was the first to recognize it. Hutton believed the Earth was older than 6,000 years, but he never said how old he thought it was. We don't even know if he gave the actual age of the Earth much thought.

Hutton was also one of the first geologists to take detailed notes in the field and describe his observations in detail. Field notes are key to understanding how you came to your conclusions. Notes are also used to confirm your observations. They are extremely helpful in the event that you cannot complete your work. Prior to him, many naturalists would just publish summaries and conclusions of their observations.

Charles Lyell

Hutton died in 1797, the same year another man named Charles Lyell was born on November 14th, in Kinnordy Scotland. Sir Charles Lyell was a lawyer by profession, but he was also a close friend of Charles Darwin. He attended Oxford in 1816 to 1821. He made his first geological observations in 1820 as he completed his first travels through rural England. He turned to geology as a full-time profession in 1823 after publishing a paper on the fossils near his home in 1822. He totally abandoned the practice of law in 1827 to pursue geology full time. He was also a devote Christian. Although he believed in an old Earth, the idea of evolution was not something he supported.

Despite his religious beliefs, he assisted Darwin in his first publication 1858. Lyell never fully accepted evolution. He did give it recognition in the 10th publication of the "Principles of Geology" in 1859. It was less of an endorsement and more of an acknowledgment of the impact of the concept of evolution on science. Although devoutly religious, he recognized the value of science and the need to keep religious preconceptions from being used to form hypotheses.

Lyell was also the first to take a guess at the true age of the Earth, based off his local observations. He believed the minimum age of the Earth was 300 million years old. He missed the modern accepted age by a lot. It is actually over 15 times older than his 300-million-year minimum age. He did not believe that an old earth contradicted his religious convictions. His estimation for the age of the Earth was built upon observations that he had made in his local area of Great Britain. He did not know of the geology on other continents. In his defense, not much was known during his time.

He was also one of the first people to try to divide the history of the Earth into ages. He coined the geologic names such as Paleozoic, Mesozoic, and Cenozoic, which we still use today, to describe slices of geologic time. His most famous work is a three volume book titled "Principles of Geology" (published from 1830 to1833), in which he coined the phrase, "the present is key to the past". This book would make the concept of uniformitarianism popular.

Charles Darwin

Lyell was also fascinated with erratic boulders that he saw across the English landscape. He requested that Darwin search for them during his voyages on the Beagle. When most people hear the name Charles Darwin (1809-1882), they think evolution and not geology. But he was a naturalist. Both life and the planet that produced it interested him. The requests made by Lyell would draw Darwin to the conclusion that glaciers were once far more widespread than they are today. He observed small out of place stones included in deposits that we consider 260 million years old today. We call these stones "drop stones". Drop stones are akin to small glacial erratics within the layers of marine deposits.

Although Darwin observed glacial processes far beyond any present glacial limits or the limits of the Quaternary glaciers at their maximum on his journeys, he never really developed the idea of ice ages. That would come a bit later. First the recent glaciations of the Quaternary would be recognized. Once those deposits were recognized it became easier to recognize other ice ages in the deep past. Any geologic observations that Darwin made were soon overshadowed by his book the "Origin of

Species" in 1859. The most famous book on evolution ever written.

The basic observations of Hutton and the expanded observations of Lyell would set the framework for geology to this day.

The geologic timescale: a dynamic concept

Although modern geology got its kick start in the 18th century, the observation that rock units exist in stratigraphic layers goes way back. No one knows for sure who first made this observation. It may have been one of those things observed over and over by hundreds or thousands of people, but never written down. Even a non geologist can walk up to a wall of naturally out cropping rock and see that it exists in layers.

Nicolas Steno

As early as the end of the 17th century, Nicolas Steno (1638-1686) was the first to state that rock layers are laid down horizontally. This would become the "law of original horizontality". Another principle he came up with is the concept that younger strata overlie the older ones beneath. This simple observation became known as the "law of superposition", a key concept in geology to this day. The present definition of the "law of superposition", simply states that "marine sedimentary rock layers are oldest on the bottom and youngest on the top, in an undisturbed sequence of rock strata". The undisturbed part means that there has not been any later folding, faulting, tectonic, metamorphic alteration, igneous intrusions, or human alteration of the rock layers. This simple observation has more than academic value. It can be used to correlate seemingly unrelated rocks in one area to another area where they have been disturbed. This can help you locate materials like ore bodies and groundwater.

Before I go further, I need to talk a bit about Steno for several reasons. His principles are often attacked by people peddling young earth creationism and flat earth. Both groups of the afore mentioned people often don't understand science and as a result they don't understand Steno's Laws. You can tell by how they attack the concepts. You cannot falsify scientific concepts by arguing loopholes. Especially ones gained through misunderstanding. Nor is the claim that something was derived 400 years ago invalidate the concept. Although I will be talking about Steno's laws in a simplistic manner herein, I encourage you to seek further reading on them, if you so wish. They are important concepts. Their misunderstanding can lead to rabbit holes of further misunderstandings of other geologic concepts.

Steno came up with more laws of stratigraphy. It's important to understand that a law in science is not the same as a law created by humans to run a society. A scientific law is a description of some phenomenon in the natural world. It can be a statement or a mathematical formula. It does not explain "why". That is in part what a scientific theory does. Scientific laws can also be applied in other disciplines as they are supposed to be universal concepts. They are not above theories.

Ok, enough about what a law is, back to Steno. He also derived "the law of lateral continuity" which states that rock units continue until they reach another solid body that stops their deposition. Rock units, or strata, do not continue forever. Eventually something will physically stop deposition. Marine deposits cannot be deposited on land. Meandering streams cannot deposit sand along their banks outside of their valleys. Evaporites cannot precipitate out of under saturated solutions. Glacial till gets deposited where glaciers once sat, not hundreds of miles outside of their furthest extent.

Paleo-topography, such as hills and islands, can serve as physical barriers to prevent deposition. This is what happened with the Cambrian marine sand deposits around the Precambrian quartzite that in part makes up the Baraboo Hills in Wisconsin. I will come back to Baraboo later. Steno's final law is my personal favorite. It is the "law of cross-cutting relationships". It states, "Rock units that cross-cut other units are younger than the rocks they cross-cut." This was an early recognition that not all rocks are simple layer cake sequences. An example would be an igneous dike that cuts across sedimentary units has to be older than those sedimentary units. This makes sense as you cannot intrude into rocks that don't exist. Igneous rocks can cut through any other rocks, this includes other igneous rocks. This can also be verified through absolute dating. If I have a dike that crosscuts a granite, the dike is younger. If I date the two and the granite comes out older, then I have a sort of checks and balance on my law. This is what we observe in the field and test in the lab. This is why Steno's laws are not just principles, as they originally were for hundreds of years until radiometric dating came into being. Steno's laws were an early attempt to derive universal principles to sedimentation that matched what was observed in the field.

Giovanni Arduino

The first real attempt to assign rock strata to named periods was first done in 1759 by an Italian named Giovanni Arduino (1714-1795). He divided the Earth into four basic geologic time periods. He named them (oldest to youngest) the primitive (or primary), secondary, tertiary, and volcanic (or Quaternary). This basic framework would only be slightly modified over the next century.

Abraham Gottlob Werner

The founder of modern German geology, Abraham Gottlob Werner (1749-1817) added "transition" between the primitive and secondary periods, thus acknowledging new data. Primitive would eventually become the super-eon Precambrian, which we now know covers 88% of geologic time.

James Dwight Dana

The Precambrian would later be divided into further eons. The Archean was the first divided out that we still use today. It was coined by a zoologist named James Dwight Dana (1813-1895) in 1872. He called it the Archaeon (later respelled to Archean) to describe the oldest era to avoid a now abandoned era called the Eozoic.

Summary of geologic time divisions

Eventually the Hadean and Proterozoic would also be divided out of the Precambrian. Secondary would be dropped and is roughly equivalent to the Paleozoic and Mesozoic Eras of today's timescale. Tertiary and Quaternary would be redefined over and over. The actual names stuck until 2013 when Tertiary was finally dropped by the International Commission on Stratigraphy. Although the Geological Society of America (GSA) did remove the term in their 2012 4.0 timescale, they reintroduce the Tertiary to their 5.0 timescale in 2018. The term Quaternary no longer applies to volcanic rocks only. It is actually tied to the modern ice age and only encompasses the most recent 2.58 million years of geologic history. It is likely only a matter of time before it is dropped as well.

Werner and Neptunism

Werner did believe in deep time; he would be known more for his failed concept of "Neptunism". He was a popular professor. Neptunism was a short-lived branch of geologic thought that postulated a global ocean existed and then receded to produce stratigraphic layers of rock. This might sound like sediment being carried in seawater to be deposited out, but that isn't exactly what Werner had in mind. It was obvious that sediment could be eroded and deposited in streams and suspended sediment was part of the concept. However, Neptunism was more a belief that the vast majority of land was precipitated directly out of seawater. Although some rocks like evaporites are deposited out of solution, rocks like shale are not. Neptunism even went as far to claim igneous plutonic rocks such as granite precipitated first from some primordial ocean. Although global sea level has risen and fallen hundreds of times in Earth's history, his model was over simplistic and was combined with Genesis in order to explain the observations. Hutton's uniformitarianism would eventually replace Neptunism as the dominant school of thought.

William Smith
In 1815 a man named William Smith (1769-1839), who was a farmer and canal builder made his observations throughout England, produced a map that would change the way we looked at the layers that make up the Earth.

Although Smith was not formally educated, he became an engineer and created the first geologic map of England based off the correlation of fossils. It wasn't until 1831 that he was acknowledged as a contributor to geology by the Geological Society of London. Although Smith's 1815 map changed our perception of the world and established the geologic mapping process, and he was aware that different fossils occur within different rocks, he made no attempt to divide the rocks into specific timeframes. I don't think this was a lack of observation on his part. It may have been the exact opposite. He recognized that subterranean forces could alter the position of strata. During this time, people were still guessing at the age of the Earth. Smith had no concept of plate tectonics, volcanic mechanisms, or fault mechanics, like we do today to aid him in geologic interpretations. He kept his map simple by correlating rocks based on fossils. Correlating strata based on the fossils within, would eventually lead to the modern discipline of "biostratigraphy".

John Philips and Adam Sedgwick

Smith's 1815 map would inspire the very first attempt to formally standardize geologic time on a global scale. In 1841 John Philips (1800-1874) would be the first to attempt to create a global standard for geologic time. Philips was an English geologist. He coined the terms Paleozoic and Mesozoic, which are still used today. Other geologists would coin the terms Cambrian, Ordovician, Silurian, Devonian, Carboniferous, Permian, Trias (now called the Triassic), Jurassic, and Cretaceous. All of which today are still used as geologic periods but have been redefined and included into the Paleozoic and Mesozoic Eras. Adam Sedgwick (1785-1873) a British geologist and priest, was another great influence on Phillip's timescale. In 1835, based on rocks in Wales, he originated the term Cambrian as a division of time. In 1840, along with Roderick Impey Murchison (1792-1871) would put to rest the "Great Devonian Controversy". This early controversy, unlike the later granite controversy, was not so much a triumph of the scientific method, as it was a compromise to settle an argument. This was a reconciliation that acknowledged two separate and the well defined fossil assemblages of the Silurian and Carboniferous, separated by a red colored rock that occurred not only in Britain but as far east as Russia as well. Thus, the Devonian Period was born. The Devonian would be established as a global geologic period and would end up being recognized globally.

Joseph Le Conte

Over time, many geologic time divisions have been suggested, such as the "Ozarkian Period" proposed by Joseph Le Conte (1823-1901)", an American geologist and naturalist. In his 1899 paper "The Ozarkian and Its Significance in Theoretical Geology" he attempted to validate the Ozarkian, but the concept came and went quickly. Conte did accomplish other things in geology. He would accompany Louis Agassiz in his 1851 exploration of Florida. Agassiz is better known for his ideas on ice ages...I will come back to him later.

Henri Becquerel and Marie Sklodowska Curie

As the 20th century approached, the geologic time scale did not change much. The time scale was only a relative way to date rocks until the discovery of radioactive decay in 1896 by Henri Becquerel (1852-1908) and Marie Sklodowska Curie (1867-1934). Radiometric, or absolute dating, would eventually confirm many outstanding concepts in geology. Such as the laws of stratigraphy. Most people think of Carbon-14 when they think of radiometric dating. But C-14 is only one of nearly two dozen methods. C-14 is basically useless for dating anything older than 50,000 years, due to its relatively short half-life of 5,730 years. To date really old rocks we use methods like Uranium-Lead, which can accurately date rocks billions of years old by using inorganic minerals like zircon.

Arthur Holmes

The very first geologic time scale to include absolute dates was published by Arthur Holmes (1890-1965), in his 1913 renowned book titled "The Age of the Earth". He came up with an age over five times older than Lyell did. He hypothesized Earth was at least 1.6 billion years old, based off the first radioactive dating techniques. This was the first laboratory conducted test ever run to answer the question of the Earth's Age. Holmes also championed Alfred Wegener's "continental drift" hypothesis, which was very unpopular in the first half of the 20th century.

Wegener would conceive one of the most revolutionary concepts ever in the field of geology. We will get to Wegener later.

Clair Cameron Patterson

The modern accepted age of the Earth was measured using radiometric dating of lead-lead isotopes by Clair Cameron Patterson (1922-1995) and George Tilton. Patterson was a geochemist born in Mitchellville Iowa. He deduced an age of the Earth at 4.55 billion years old in 1956. This age has essentially remained unchallenged to this day. His methods not only helped derive the age of the Earth, but made us realize the dangers of lead poisoning.

Modern timescales

It wasn't until 1977 that scientists began to internationally collaborate on the standardization of a geologic time scale under the "Global Commission on Stratigraphy" (now called the "International Commission on Stratigraphy"). Since then, the geologic timescale as been divided and the ages changed based on better dating techniques. In every revision the margin of error decreases. It still isn't "perfect" and never will be. Perfection is a useless abstract concept in science as is absolute truth. It has been a long road to understanding the age and time divisions of the Earth, a process that has taken over 300years!

Dividing the Earth's history into named units of time was one thing. Understanding how the surface of the Earth developed over time was a totally different beast.

Continents moving around the globe: Yeah. Right?

Until the 1960s no hypothesis in geology has been as highly scorned as the concept of "continental drift", which was eventually abandoned in favor of the theory of "plate tectonics". Continental drift simply states that all the continents were once joined together and subsequently moved to their present positions by moving upon the ocean floor. The theory of plate tectonics is perhaps the most complex theory developed in geology and cannot be attributed to one person being responsible for most of it.

I need to make a clarification here. Most of you have heard of continental drift. It is not the same as plate tectonics, although people talk about them as if they are synonyms. Continental drift was the pioneering hypothesis that would have to wait until the technology was developed to look beneath the oceans and the core concept that continents move would be a part of plate tectonics, but continental drift is not the framework for plate tectonics. Plate tectonics was developed in the 1960s and 70s, based on initial observations mostly made in the 1940s and 50s. Beginning in the 1970s and persisting through the 80s the theory was developed based on additional field observations and focused measurements. By the 1990s a firm understanding of how the plates move about the surface of Earth was out, and it hasn't changed much to this day.

The idea of continents moving around the globe is not new. It is an old idea used to explain the appearance and isolation of animals on different parts of the globe. As early as the year 1596, the cartographer Abraham Ortelius was the first to observe that the continents look as if they do fit together, like a jigsaw puzzle along the Atlantic coastlines.

Alfred Lothar Wegener

It wasn't until Alfred Lothar Wegener (1880-1930) came up with the hypothesis of continental drift in his 1915 book "The Origins of Continents and Oceans". Wegener was born in Berlin. He was a meteorologist by profession but loved geology. He took part in four polar expeditions throughout his life, mostly to find evidence for continental drift. His fourth expedition would cost him his life and bury his journal under the snow. His journal would never be found.

His idea of continental drift was scorned at the time, and later demonstrated to be correct in its basic premise. Like other people, Wegener noticed that the continental coastlines along the Atlantic Ocean appear to "fit together". He also noticed similar fossils were found on lands now separated by thousands of miles of ocean. He postulated all the continents were once joined in a supercontinent called Pangaea.

Although the idea seemed sound to him, he provided no real mechanism for how continents could move about the globe across the ocean floor. He thought tidal motion may have been responsible. Even in the early 20th century it was well known that tidal forces acting on the Earth by the moon and sun, were not strong enough to move continents. So, his idea was met with skepticism and ridicule, and rightfully so. After all, "Extraordinary claims require extraordinary evidence." Carl Sagan (1934-1996). In geology, and science in general, if you have a revolutionary claim about how Earth processes occur, you better have an observable and/or testable mechanism, with lots of evidence to back it up.

Wegener's hypothesis came out at the same time that the first highly detailed and accurate maps of the world were produced. The first accurate map depicting all the continents (except Antarctica) and their coastlines was released in Germany in 1870 by C.F. Baur and T. Bromme titled "Latest Map of the Earth", which was a Mercator projection.

With modern maps in hand, it was easy to see how most coastlines fit together. One or two could just be a coincidence, but most of the Atlantic coastlines almost fit perfectly, with some minor exceptions. Since in science correlation doesn't necessarily equal causation, Wegener's idea would have to wait.

At this time, the coastlines were accurately located, but the ocean floor was still a mystery. No one had attempted to map the ocean floor until the 1920s.

Arthur Holmes and the mantle

Let's go back to Arthur Holmes for a moment, the guy who first published an age of the Earth using radiometric dates. In 1929 he came up with idea that the interior of the Earth is driven by convection cells in the mantle through the processes of "thermal convection". Basically, the same concept when you boil water on a hot stove. The hot material rises as it's heated from the bottom and sinks back down when it reaches the surface and cools. He postulated that if the convection force were strong enough, this could move the continents apart. Today we know he was partially correct. Alas, there was no evidence at the time to back-up his idea. It wasn't until we started to map the ocean floor that the fuzzy picture's resolution was increased.

Holmes had no firm idea as to what the internal composition or density of the Earth was all the way to its core. The mantle, the layer beneath the crust, is actually solid and not liquid. It does undergo convection, but very slowly. No where near fast enough to be the only driving force behind continental drift. The mantle behaves in a ductile manner, instead of brittle like most of the Earth's crust. The crust is the part of the Earth that we live upon as well as forming the floors of the ocean. It is also the thinnest layer of the Earth. Varying from the ocean crust being roughly 5 miles to the continental crust being about 30 miles thick. Seismology (earthquake detection) had been used since the turn of the 20th century.

Andrija Mohorovičić

It wasn't until 1937, using new techniques, that we would begin to understand the interior of the Earth all the way to the core. Early on, one of the founders of modern seismology was the Croation geophysicist Andrija Mohorovičić (1857-1936). In 1909 he discovered the border between the crust and mantle. It would become known as the Mohorovičić discontinuity or the Moho for short.

Land bridges

What was the accepted theory of how extinct animals appeared on different continents in the early 20th century? Up until the 1960s the explanations for extinct and living animals traversing the globe was through random land bridges that would emerge from the sea floor and then become submerged again. Land bridges can occur, like during an ice age. For instance, when glaciers covered the northern part of North America, sea level dropped enough to close the Bering Straight and create a temporary land bridge between Siberia and Alaska. There is a far cry from lowing sea level enough to close a 50-mile wide straight and lowering it enough to connect continents thousands of miles apart. The oceans never retreat enough to connect South America and Africa. The hypothetical land bridges needed their own driving force to be raised and lowered from the ocean floor. No viable mechanism for these land bridges were ever really devised either. It was just accepted dogma at the time and Wegener's idea rocked the boat.

George Gaylord Simpson

The idea of land bridges had begun to fall out of favor in the mid 20th century. Especially in 1940 with George Gaylord Simpson's (1902-1983) publication called "Mammals and Land Bridges". Simpson was an American paleontologist, and he was perhaps the most influential paleontologist of the 20th century. Simpson set out to catalog the various theoretical bridges from one continent to another. His work focused mostly on recently extinct and living mammals. It soon became apparent to him that the amount of data and work needed to match the observations was almost beyond human capability. This planted the seed of doubt in the minds of other geologists. If he couldn't even propose a simple mechanism to match the observations, how was anyone going to do it with animals that went extinct 65, 150, or 300 million years ago?

The advent of sonar in the 1920s revealed that there was topography on the ocean floor, but the picture would remain fuzzy for the next couple of decades. In 1950 linear mountain ranges were discovered on the ocean floor, dubbed "mid-ocean ridges". Prior to this, the ocean floor had been thought of as flat and featureless, minus the occasional bumpy spots picked out by pre-World War II sonar. The discovery of mid-ocean ridges would revolutionize our concept on how the Earth is shaped and made us give the idea of continental drift a second look.

Harry Hammond Hess

In 1962, Harry Hammond Hess (1906-1969) a WWII American navy officer and geologist, published his paper the "History of Ocean Basins". In which he suggested that magma rises up from the interior along the mid-ocean ridges, slowly driving the continents apart and in deep ocean trenches, the ocean floor is being recycled back into the interior. This is a far more viable idea and was far different from Wegener's belief that the continents plow across the surface of a featureless ocean floor.

As it turns out, the continents are situated within the ocean floor and are just along for the ride as the ocean expands in one area (along mid-ocean ridges) and gets returned to the mantle in another (deep ocean trenches). This became known as Hess's theory of "sea floor spreading", this still wasn't enough to demonstrate that continental drift, or something like it, was indeed occurring.

Fred Vine and Drummond Matthews

While Hess was developing his concept, the U.S. Navy was mapping the magnetism of the rocks on the sea floor. In 1963 it was noted that bands of polar reversals were recorded in the rocks on the sea floor by Fred Vine (1939-?) and Drummond Matthews (1931-1997). Although the Earth's magnetic field has been known since ancient times, it wasn't until the 20th century that we not only realized magnetic north isn't fixed, but occasionally will reverse itself. At some points in the geologic past, a compass will point north at others, it will point south.

Magnetic reversals on the ocean floor

No human being has witnessed a magnetic reversal, but we know they occur because of minerals in volcanic rocks. As magma and lava cools the iron minerals, such as magnetite, will align themselves with polar north. Some of the rocks on the ocean floor indicate a polar north pointing south, then north again, then south, then north, and so on. Thus, the theory of plate tectonics was born. The Earth's surface is "cracked" and can be divided into a dozen major tectonic plates. The plates are bounded by the "cracks" along mid-ocean ridges and the deep ocean trenches. The continents sit within these plates, not on top.

Since its conception, the theory of continents moving towards their present position from somewhere else has evolved. Today, we can actually measure the rate of sea floor spreading with satellites and by bouncing lasers off the moon. The continents are indeed moving apart along the Atlantic Ocean, at a rate of about how fast a human fingernail grows.

Holmes was on to something when he theorized convection cells as one of the driving forces moving the continents. Hess's theory of sea floor spreading, or ridge push, explains the creation of new ocean floor and the destruction of old ocean basins in subduction zones. Subduction zones tend to be located where the ocean crust is the oldest. So, the main mechanism of plate tectonics has been worked out? Not so fast.

Mechanisms that drive plate tectonics

As the ocean floor ages, it becomes thicker and denser. It then begins to break and sink under its own weight due to gravity. Eventually slab pull encourages the old ocean floor to subduct under more buoyant younger crust. Although ridge push is one of the forces driving plate tectonics, slab pull appears to be the dominant driving force, with mantle convection being the third mechanism.

Slab pull was first proposed as a mechanism for plate tectonics by Donald Forsyth (1949-?) and Seiya Uyeda (1929-?) in 1975 when they published the paper "On the relative importance of the driving forces of plate motion". This is why subduction zones are located where they are, along ocean trenches where the oldest and thickest ocean crust tends to be located. If you plot the location of volcanoes and earthquakes on a map, they tend to follow the subduction zones, where the ocean crust is returning to the mantle. Friction causes some rock to melt as one plate dives beneath another. This causes magma to rise to the surface and build volcanos.

Ocean crust v. continental crust

The oldest ocean crust dates back to the time just before the dinosaurs. It is about 270 million years old and is in one small part of the Mediterranean basin. Yet the continents date back billions of years. How is this possible? The continents are less buoyant than the ocean crust. When two continents physically collide, as in the case of India and Asia, one does not subduct under the other, as ocean crust does. The process instead builds mountains until it stops, and subduction begins elsewhere.

The process of plate tectonics has been moving the continents around the globe and has been active as we know it today, for about 2.5 billion years, perhaps as far back as 3.2 billion years. Most people are aware of the supercontinent Pangaea. However, there have been at least two others and possibly as many as five throughout the billions of years that Earth has been orbiting the sun.

Plate tectonics answers so many questions that were unanswered in geology. For example, the building of mountain ranges, the presence of marine fossils in land locked areas, the distribution pattern of fossils, the creation of faults, the existence of volcanoes along the margins of continents, and the apparent polar wandering of Earth's magnetic field.

The discovery of mountain building

Plate tectonics answers the question of why mountain ranges form, but the mechanics of their formation was figured out much earlier.

Structural geology
Today we call the division of geology that deals with the three-dimensional distribution of rocks based on their geologic histories as "structural geology". Structural geology is a key component to making geologic maps and cross sections, which in turn can be used to predict the locations of mineral resources at the surface and deep underground. If you have ever flown over the American Appalachian Mountains, you will notice that some of them appear to form large "V" structures. These are plunging folds. A fold occurs where the originally deposited horizontal rocks have been tilted and folded into a new position by tectonic forces. Structural geologists measure the layers in those folds to compute their extent, and to figure out the history of the rocks after they were deposited.

Surprisingly, the relationship of rock layers in mountain ranges was figured out about 50 years before the theory of plate tectonics. That isn't as surprising as where the history of mountain ranges unfolded. You may think that the structural features of mountains were deciphered in an area with a lot of mountains, like the Appalachians, or Rockies, or Himalayas, or the Alps, but it wasn't. It was figured out in Wisconsin, specifically in the Baraboo Range, in the south-central part of the state.

The Baraboo Range

The Baraboo Range is a large geologic structure referred to as a syncline, that is located in south central Wisconsin. The rocks are mostly made of an extremely hard purple rock called "quartzite" (the metamorphic version of sandstone). The rocks are the remnants of an ancient mountain range that existed in a roughly east-west line across the southern part of Wisconsin about 1.6 billion years ago. The mountains have long since been eroded and the Baraboo Range represents a small preserved fossil of that ancient mountain range. The Baraboo area is also a teaching ground for geologists. Every geology major in the Midwest has spent time at Baraboo, including myself. The Baraboo Range displays many geologic structures throughout its area. It is an excellent place to see the power of tectonics in action without having to cover vast distances or head into difficult to access areas.

When the Baraboo Range first came to the attention of geologists in the late 19th century, its exact relationship to the flat surrounding rocks was a complete mystery. The rock units are tilted, some vertically. Yet the rocks that surround the range are horizontal.

Charles Richard Van Hise and Charles Kenneth Leith

The mystery remained until two men figured it out. The more prominent of the two was Charles Richard Van Hise (1857-1918), born in Fullerton Wisconsin, he would later become the president of the University of Wisconsin from 1909 until his death. He worked closely with the United States Geological Survey (USGS), which was founded in 1879. Van Hise produced many publications that were firsts. In 1904 he published the "Treatise of Metamorphism", the first comprehensive writings on how rocks become metamorphosed. He also published the first Precambrian rock distribution map of North America in 1909, along with Charles Kenneth Leith (1875-1956). Two years later, Van Hise published the first geologic writings on the Lake Superior region in 1911.

Van Hise is one of the rare geologists who has a plaque on a protected monument, dedicated to him on one of the very rocks in where he figured out the mystery of the Baraboo Range. It was at Van Hise Rock (and other locations in the Baraboo Range) where he figured out "stress fields". Rocks on a large scale reveal their history within on a smaller scale. By measuring these small-scale geologic features, you can figure out the bigger picture on a regional scale.

He correctly determined that the Baraboo Range was a syncline and not an anticline. A syncline is a large somewhat bowl shaped geologic structure. An anticline is the opposite, an upside down bowl, although the Baraboo Range is more spoon shaped, with one long side bent inwards. Drilling in the middle of the syncline would later confirm that Van Hise was correct. This is just one way in which geology can make testable predictions.

Charles Leith worked closely with Van Hise in the field and as a coauthor. He would become the geology department chair at the University of Wisconsin, a position he held for 31 years. He would also go on to receive the Penrose Medal in 1942, one of the highest honors in the earth sciences.

It wouldn't be until after the theory of plate tectonics that the first detailed geologic map of the Baraboo Range was compiled in 1970 and is still used today.

Wisconsin also holds a bit of geologic history in another way. General, and later President, Ulysses S. Grant's (1822-1885) tomb was quarried from granite. Red granite was established as the state rock of Wisconsin in 1971. His tomb was quarried in Montello, Wisconsin. Granite rocks are well exposed in southern and northern Wisconsin. However, the origin of granite was hotly debated, even as Van Hise figured out the dynamics of mountain ranges.

Great granite debate: from water to magma

Granites are one of the most common igneous rocks on the planet. They are a group of rocks that are so coarse grained that individual crystals of feldspars, quartz, and dark minerals can be seen without magnification. They make up the core of continents and are all but absent in the ocean floor. Through modern techniques, we can melt rocks in the lab and let them cool at different rates to observe which minerals and rock types ultimately form. This experimentation would demonstrate how granite forms. We now know granite forms from slow cooling magma deep underground. This understanding wasn't always the case. There was an argument that raged from about the 1880s to 1958 called the "great granite debate or controversy".

Werner and Neptunism

Remember Abraham Gottlob Werner? He didn't only devise one of the first geologic time scales, he was also responsible for what would be called Neptunism. It is the concept that the Earth was covered in a vast ocean that receded over millions of years. The theory is simple, too simple. It states that there was once a vast ocean covering the Earth. In this concept the continents did not exist in the beginning, and all the minerals were dissolved within the waters of the ocean. Slowly the heavy minerals settled out and formed granite, gneiss, and schist which sank to the sea floor and formed the core of the continents. Gradually the less dense rocks like sandstone, limestone, and shale precipitated out to cover the granite and finally the continents emerged from the ocean as it receded. No one could explain where the water in this primordial ocean went as it became shallower and formed the continents. That was just one of the many problems with Neptunism.

Of course, there was no way to test Neptunism in the 19th century and other geologists like Hutton, had already correctly surmised that coarse grained igneous rocks were the product of cooling magma, a concept known as Plutonism.

Hutton, unconformities, Neptunism, and metamorphic rocks
Hutton also noticed that some rocks appeared out of sequence and did not conform to the law of superposition. These rocks were often separated by missing strata, which became known as "unconformities". Unconformities are the contact where two or more rocks of different ages are separated by a volume of missing time. There are many types of unconformities. Some are caused by erosion, landslides, building of mountains, or by an intruding body of magma. The amount of missing time can be small or exceptionally large. The Silurian rocks in Illinois are separated by three major unconformities that represent several hundred thousand to a couple of million years of missing time. A couple million of years is a lot to a human being but is only 0.000004% the age of the Earth…a drop in the bucket.

You have unconformities like the one exposed along Lake Superior in Marquette Michigan, where two Precambrian rock units, the Mesnard Quartzite and the Jacobsville Group, are separated by an erosional unconformity of almost 1.4 billion years. That is slightly more than 30% the age of the Earth. Although Hutton recognized unconformities, he had no idea of the time scales involved, and there was no way to test this at the time either, because we had no way to directly date rocks in the 19th century.

So, the debate continued essentially until the advent of radiometric dating. It was soon realized that some rocks near one another in highly deformed areas, like mountain ranges, had highly variable ages. If all of the world's granites settled out of a primordial ocean first, then they should all be layered and older at the base and younger near the top. That was not what we observed, and Neptunism finally faded into history.
As in the case of most things in the natural world, neither Plutonism nor Uniformitarianism were the sole shapers of the Earth, neither concept was completely wrong. Even aspects of Neptunism weren't totally out of reality. Other forces such as plate tectonics and the role of impacts, had not even been conceived yet.

Some rocks do precipitate out of sea water, such as certain limestone, evaporites, and banded iron formations. These are called "chemical rocks". Rocks such as sandstones, conglomerates, and shales are formed from the erosion of older rocks and deposited elsewhere and are referred to as "clastic rocks". Some rocks like coal and black shale are deposited by the dead bodies of once living organisms, so are some limestones. Other rocks like granite, basalt, and rhyolite form from the cooling of magma and lava. Then you have metamorphic rocks.

It was noted in the early 19th century that some rocks were so highly altered that their original structure and initial rock type have been converted into an entirely different rock type. These rocks became known as "metamorphic rocks". Common metamorphic rocks are gneiss, schist, marble, and quartzite.

Until the advent of laboratory experimentation, granite was often considered metamorphic. Metamorphic rocks are one of the three main types of rock, along with sedimentary and igneous. They form when a rock undergoes extreme heat and pressure, but not to the melting point. Magma and lava are not rock. Rocks are essentially solids that are an aggregate of one or more minerals. Minerals are made of crystals, which are always in a solid state.

After the death of Neptunism, granites still posed a mystery. It was clear they did not precipitate from sea water. Their chemical composition is different from what would be suspected to well up from the interior of the Earth, or so it was thought. Today we know that granites cool very slowly underground. The magma they form from often melts the surrounding rock and incorporates it into the magma as it cools. Granites don't punch their way quickly through the Earth's crust like fine grained basalts and rhyolites do in violent eruptions. The result is they tend to be made of less dense minerals as they incorporate the surrounding rock into their magma.

Bowen and Tuttle

Until experiments were run by Norman Levi Bowen (1887-1956), a Canadian geologist born in Kingston Ontario, the debate as to the origin of granite, and if it was an igneous or metamorphic rock, was an open question. In 1928 he published "The Evolution of Igneous Rocks", in which he revolutionized petrology. Petrology focuses on the composition and microscopic structure of rocks. It is not only observational based geology but also experimentally based.

Through his experiments, he learned that most minerals do indeed form from the cooling of magma or lava. By melting rocks and cooling them at different rates and under different pressures, he could watch the minerals form, this led to the

"Bowen's Reaction Series". Certain minerals always form with other minerals. Other minerals never form with certain other minerals. All igneous rocks are made of minerals, so the experiments demonstrated that granite forms slowly from deeply buried magma, with a specific chemical composition, and not through the process of metamorphism.

Bowen worked closely with an American petrologist, a man named Orville Frank Tuttle (1916-1983), from 1948 up until is death in 1956. Tuttle went by Frank and in 1958, he published the 153-page paper titled "Origin of granite in the light of experimental studies in the system $NaAlSi_3O_8$-$KAlSi_3O_8$-SiO_2-H_2O". Bowen is on the paper, even though it was published two years after his death. Although Tuttle published it, Bowen developed the concepts and set up most of the experiments Tuttle ran and interpreted, making him instrumental to the paper. The paper is a very extensive and thoroughly documents the results of the experiments run. It was also one of the first papers to include the presentation of data in ternary diagrams. Ternary plots are equilateral triangular diagrams used to plot ratios of three items instead of two.

The 1958 paper was a triumph of the scientific method over just arguing and talking about things. The paper demonstrated through experimentation that can be reproduced, granitic rocks were almost exclusively magmatic in origin.

It put to rest the great granite debate, but another far colder mechanism also helped shape the Earth's surface.

Ice ages: a cold concept

Pierre Martel

During the time of Hutton and Lyell, another great shaper of the Earth was coming to be realized, the power of ice. Beginning in the 18th century many individuals were told tales, by natives, of how glaciers once extended further than they do today. In 1742 the engineer Pierre Martel (1706-1767) was told stories from the locals in the Alps, about how the glaciers once extended further than they do and that the large out of place boulders, known as erratics, were left by the glaciers.

Jean-Pierre Perraudin and Jean de Charpentier

Others such a Jean-Pierre Perraudin (1767-1858) and Jean de Charpentier (1786-1855) also wrote on the subject of past alpine glaciers. The fact erratics seen far from any existing glaciers, such as the ones on the British Isles, where there were no present glaciers, remained a curious enigma.

The idea of ice as a major sculptor of the landscape took awhile to get going. Unlike catastrophes caused by floods and other acts of God, the bible mentions nothing about ice. Although the bible is not a scientific document, the stories in Genesis were still held by many of these early pioneers of geology. For some it was rigid belief, but for others, it was for lack of a better explanation. So, the idea of massive glaciers covering the continents was slow to develop.

Daniel Tilas

Scandinavian evidence for ice ages was noticed early on. Daniel Tilas (1712-1772) a mining expert from Sweden first suggested that erratics were transported by icebergs and dropped into the sea as they melted. Hutton believed the erratics in the Alps were left by long gone glaciers.

Louis Agassiz

Lyell even asked Darwin to look for and catalog any erratics he may have encountered on his journey aboard the Beagle. A man by the name of Louis Agassiz (1807-1873) also influenced Darwin. The two men began correspondence in the 1840s. Agassiz was a Swiss naturalist who believed strongly in creationism but did not believe in a literal interpretation of the bible. He believed there was a third creation event. He thought that sometime after the great flood, there was a worldwide deep freeze. He went to the Amazon to seek evidence of the global ice age but found none.

Jens Esmark and Milutin Milanković

Although Agassiz took global ice coverage to the extreme, he wasn't the first to propose that glaciers once extended past high mountain ranges and Polar Regions. Jens Esmark (1762-1839) in 1824 published his paper "Contribution to the History of the Earth", where he proposed worldwide ice ages. Esmark also was the first to propose that climate change and variations in Earth's orbit may have been the driving force behind the advance and retreat of glaciers. This concept would further be developed by Milutin Milanković (1879-1958) a Serbian mathematician and geophysicist, and founder of climatology. He would be responsible for formulating the predictive cycles that would cause Earth's orbit and relationship to the ecliptic to change over time. These fluctuations that he derived in the 1920s, would become known as the Milankovitch Cycles.

Erratics, till, and striations

Glacial erratics may have ignited the debate on the possibility of ice ages but are not the only evidence. While most proponents of a lost age of ice were tramping around Europe looking for evidence, the Americas were being settled and geologists began to notice something covering the Midwest of North America. Thick deposits of jumbled clay, silt, and sand were noticed. These unsorted deposits would often contain large erratics within them. These deposits became known as till. Till in the Midwest can be up to 500 feet thick. This was far thicker than till deposits in Europe. Often noticed at sharp unconformities between till and bedrock, long linear grooves were observed. These became known as "striations". Striations occur when a glacier drags pebbles along its base and cuts into the hard underlying bedrock. This discovery of till led to the immediate attempt to classify it.

Archibald Geikie
Archibald Geikie (1835-1924), a Scottish geologist, discovered plant remains in till and proposed that maybe the glaciers had advanced and retreated many times.

Till and its origins
Several origins for till deposits were suggested. Of course, the great biblical flood was initially used by some to explain the unsorted material. There are no surface till deposits, from the Quaternary ice ages below 37°N latitude. Once their limited extent was realized, a sea advance (or transgression) from the north was suggested. No known marine deposits would leave such thick unsorted deposits hundreds of miles inland in the manner in which they were observed in the field. In addition, no marine fossils were ever found in the till. Although Neptunism was dead by the 1890s, sea transgressions and regressions did play an important role in depositing sediments. Sea advance and retreat wasn't the explanation either.

Willard Libby
With the discovery of carbon-14 dating in the late 1940s by Willard Libby (1908-1980), we were able to date the organics within glacial till and associated glacial deposits. Archibald

Geikie may have been the first to suggest smaller ice advances and retreats within a larger ice age, it really wasn't a viable concept until the advent of carbon dating. C-14 is extremely useful, but it is limited in its use. Although extremely accurate, it is unreliable on dead organisms older than 50,000 years. The present ice ages extend back 2.58 million years. This time span is the Quaternary Period. The ice advanced and retreated south of the 50°N parallel at least five times. For the correlation of older till deposits we strongly rely on fossils and other dating techniques.

Earth's ice, or lack of ice, past

Although the most recent Quaternary glaciers are the modern focus of glacial action outside of alpine and polar areas, the Earth has been glaciated before. The oldest known till deposits in the world date back 2.4 billion years and are well exposed in southern Ontario and the Upper Peninsula of Michigan. There have been at least four major ice ages since then.

Although ice ages date back to the beginning of plate tectonics, Earth is usually an ice free planet. The poles began to most recently ice over about 20 million years ago, but before that the Earth seems to be totally ice free up until about 320-260 million years ago, during the Carboniferous and Permian Periods. There isn't even any evidence of any polar ice caps between 260 and

20 million years ago. This isn't to say it never snowed. There just aren't any deposits left by glaciers from that time.

Ice ages are exceedingly rare in the history of the Earth. So rare that if you combine all the time when Earth had either ice caps or an ice age it would only account for about 3% to 5% of Earth's entire history. If we did not have ice caps today, I wonder if we ever would have been able to prove that erratics, striations, and till were deposited by long dead glaciers that spread over the Northern Hemisphere or if their existence would still be debated today.

Planetary geology: a new frontier

The advent of space exploration in the last half of the 20th century has greatly expanded our knowledge of geologic processes, not only on other worlds but here on Earth.

The Apollo moon missions (1969-1972) brought back rocks that not only showed the Earth and moon are close to the same age but also helped develop the leading theory that the Moon formed from a collision of a large Mars sized planet over four and a half billion years ago.

What we have learned from both manned and robotic missions is that the processes that shaped the Earth are mostly different from those that shaped other bodies in the solar system. Planetary geology has only been around for about 50 years and is still a young branch of geology.

There is still plenty of unknown geologic processes at work outside the Earth, that we have only begun to understand. It appears that all bodies in the solar system formed together from a collapsing cloud of gas and dust, yet each one has its own unique history and evolution just waiting to be discovered.

Unsolved mysteries in geology

The geology and the evolution of other worlds are not the only unsolved mysteries in geology. There are plenty of unanswered questions, right here on Earth. I am going to leave you with a short list of the many unanswered questions in geology. I figure it's a good way to make you ponder upon all that we do not know.

1) What is the exact nature of the magma chambers that underlie Yellowstone National Park, and is it possible to predict their behavior?
2) We are most likely in an inter-glacial period of an ice age. Will there be another advance of the ice and how has human civilization affected the next advance of ice, if any?
3) Were the late Precambrian ice ages really a global phenomenon as the "snowball Earth" hypothesis suggests?
4) Did all the planets in the solar system form at the same time or are some possibly slightly older? Was there a "proto-solar system" before the one we observe today?
5) Prior to 3.2-2.5 billion years ago, plate tectonics did not operate as it does today, if at all. What other

process was dominant before then? When will plate tectonics end?

6) Is it possible to accurately predict earthquakes and volcanic eruptions?

7) Will the Atlantic Ocean continue to widen millions of years into the future, or will the East Pacific Rise close the Atlantic Ocean?

8) How long does the Earth actually have before it becomes uninhabitable?

9) When will the Earth's magnetic field switch from north to south? Is it happening right now? How long will it take?

10) Did the Earth always have a solid core?

11) How did the first life on Earth appear and where?

12) How will the never ending, always increasing, human population affect the processes that not only affect life, but the physical processes that shape the Earth?

13) Did the 1.85 billion year old Sudbury impact cause the first mass extinction among single-celled life forms?

Glossary of scientific terms

Below are a handful of definitions have a different meaning in scientific usage than they do in lay speak. It is particularly important to understand this. Way too often people who try to falsify scientific concepts attack it by using colloquial definitions where they do not apply. Of course, you cannot falsify any scientific concept in this manner. That game may work in politics and the courtroom, but it has no place in the scientific method. In order to falsify a scientific concept, you need to collect your own data, and form your own testable hypothesis. Saying things like, "evolution is only a theory", not only shows your ignorance of evolution, but also exposes your ignorance of science.

Analysis: The act of comparing the results of an experiment or observations to the prediction(s) put forth in the hypothesis.

Conclusion: A summary statement of results that either support or contradict a hypothesis.

Data: A piece of verifiable information gathered through observation and/or experimentation.

Demonstration: A model used to illustrate some part of a well supported scientific concept, in order to simplify and teach some principle. A control is not needed.

Evidence / empirical evidence: Factual information gathered through observation, investigation, or experimentation that is useful in developing a conclusion based on analysis.

Empirical evidence remains the same no matter who collected it. For example: If I draw a disk-shaped object, I have evidence that I made a drawing, not evidence of alien visitation or even of frisbees existing.

Experiment: A test under controlled conditions made to record observations and/or data, in order to examine the validity of a hypothesis. A control is used.

Fact: Objective and verifiable observation or data. Facts are not explanations. Facts are used to develop a hypothesis or theory.

Facts are not "above" theories. Some scientists will use the word "fact" to mean theory. I don't like that application. It muddies the waters of an already misunderstood definition.

Falsify / falsifiable: The ability to show that a scientific concept, usually a hypothesis or theory, is incorrect utilizing the scientific method. Often synonymous with refutable.

All scientific concepts have the inherent ability to be falsified. If a concept cannot be falsified, it may be a philosophical concept, but it is definitely not a scientific one.

Geology: The study of the physical processes that shape planetary bodies.

Hypothesis: A limited concept or idea that is meant to explain some phenomenon that is testable through experimentation and/or observation, that will lead to further investigation. Any further investigation with either support or falsify/refute the hypothesis.

Law: A statement or expression based on repeated and verified experiments and observations, that describe a targeted aspect of the natural world. Laws can be mathematical or written statements.

They are often applicable across disciplines and are applicable to the natural world as a whole. The laws of physics are not only relevant in physics, but they matter in chemistry, geology, and biology. Laws are not above theories. They are used in the development and application of theories.

Model: A systematic description or illustration of an object or phenomenon that has characteristics of said object or phenomenon that is based on observation and/or data. They can be visual, mathematical, or material. They are often used to aid in predictions set forth by a hypothesis. Models can be used to demonstrate well established scientific concepts as well as to modify existing concepts in the light of new data.

Most models are not meant to be 100% reflective of reality, just some aspect of it. Material or visual model do not have to be "to scale". By nature, mathematical models are not to scale.

Natural world: The physical universe in which we live, where observable phenomenon occur that follow the laws of physics.

Observation: A statement of knowledge gained through either the senses or calibrated scientific equipment. Once an observation is verified it becomes a fact/data and can be used as evidence to support a statement.

Prediction: An accurate and precise forecast of some future situation derived from experimentation and observations in order to support a hypothesis.

The subsequent failure of a prediction can be a falsification of a hypothesis. Vague statements are not predictions. For example, "Some time in the future you will trip and fall", is not a scientific prediction. "On the 17th of this month, at 17:15$\pm$00:07 the mailman will walk down my sidewalk and trip on the 5th concrete block south of the driveway.", is a prediction, especially if derived from observations and data.

Question: An unbiased inquisitive statement to address an unsolved problem or issue that leads to the desire to find an objective or testable answer.

Science: A methodological descriptive process that follows the scientific method and attempts to describe some aspect of the natural world through testable means.

Scientific method: A systematic procedure consisting of observations, experimentation, the collection of fact and data; in order to form, test, modify, or falsify a hypothesis. It is not a rigid process and may vary slightly. It usually starts with a question or basic observation. After that research and further observations are made to form a hypothesis. Then the hypothesis is tested, usually through experimentation but it can be through other methods in order to collect data. The data is then analyzed and the conclusions, whether favorable or not, are drawn from the analysis.

Theory: A well supported explanation of some aspect of the natural world, based on a body of knowledge that has been repeatedly confirmed/verified through observation and experimentation, despite all attempts to falsify it through the scientific method.

Theories are our best working explanation of some phenomenon using the currently available evidence and can be modified with new data. In some cases it can be rejected. Falsifying a theory isn't as simple as falsifying a hypothesis. Hypotheses are targeted, theories are encompassing. A theory is NOT a guess.

About the author

Steven Baumann is a licensed professional geologist who graduated in 2001 from Illinois State University. He received his professional geologist (P.G.) license in 2011. Steve is the author of several roadside geology books and journal articles.

Steve presently works on the Precambrian rocks in the Lake Superior Region, mostly in the Upper Peninsula and Ontario. As well as the glacial and Paleozoic bedrock in Illinois, Indiana, and Iowa.

In 2009 he founded the "Midwest Institute for Geosciences and Engineering" (MIGE), a non-for-profit organization that explores the geology of the Midwest.

www.mige-web.org

Acknowledgments

There have been many people over the years who have encouraged my interest in the earth sciences. I thank you all!

Special thanks to my wife Sarah M.H. Baumann
And others: Dr. David H. Malone, Dr. Elisa J. Piispa, Sandra K. Dylka, Mary J. Pryjda, Michelle J. Abrams, Kat Hatziavramidis, and Ronnie George.

References

Baur, C.F., Bromme, T., 1870, *Latest Map of the Earth*, Mercator projection map

Bowen, N.L., Schairer, J.F., 1928, *The Evolution of Igneous Rocks*, Dover Publications

Cohen, K.M., Finney, S., Gibbard, P.L. (drafters), 2013, *International Chronostratigraphic Chart*, International Commission on Stratigraphy, v2013/01

Conte, J.L., 1899, *The Ozarkian and Its Significance in Theoretical Geology*, The Journal of Geology, Vol. 7, No. 6

Dalziel, I.W.D., Dott, R.H., 1970, *Geology of the Baraboo District Wisconsin*, Geological and Natural History Survey, Wisconsin, Information Circular Number 14

Dana, J.D. and Sillman, B., 1872, American Journal of Sciences and Art, Series 3, Volume 3. Dana, J.D., *Green Mountain Geology*, p.250-256

Forsyth, D. and Uyeda S., 1975, *On the relative importance of the driving forces of plate motions*, Geophysical Journal International, v.43, i.1, p.163-200

Hess, H.H., 1962, *History of Ocean Basins*, Princeton University, Princeton, N.J., Petrologic Studies: a volume to honor, pp. 599-620

Holmes, A., 1913, *The Age of the Earth*, Harper and Brothers, London and New York (publisher)

Imbrie, J., Imbrie, K.P., 1986, *Ice Ages: Solving the Mystery*, Harvard University Press

Lyell, C., 1830-1833, *Principles of Geology*, Penguin Classics, Abridged Edition, June 1, 1998

McIntyre, D.B., 2012, *James Hutton: The Founder of Modern Geology*, 2nd Edition, National Museum of Scotland

Rudwick, M.J.S., 1985, *The Great Devonian Controversy*, Science and its conceptual foundation series, ISBN: 9780226731025

Simpson, G.G., 1940, *Mammals and Land Bridges*, Journal of Washington Academy of Sciences, 30 (1940), pp. 137-163

Mathez, E.A. (editor), 2001, *Earth: Inside and Out*, American Museum of Natural History

Smith, W., 1815, *A New Geologic Map of England and Wales*

Tuttle, O.F. and Bowen, N.L., 1958, *Origin of granite in the light of experimental studies in the system $NaAlSi_3O_8$-$KAlSi_3O_8$-SiO_2-H_2O*, The Geological Society of America, Memoir 74

Van Hise, C.R., 1904, *Treatise of Metamorphism*, United States Geological Survey

Van Hise, C.R., 1909, *Pre-Cambrian Geology of North America*, United States Geological Survey, Bulletin 360

Van Hise, C.R., Leith, C.K., 1911, *The Geology of the Lake Superior Region*, United States Geological Survey

Walker, J.D., Geissman, J.W., Bowring, S.A., and Babcock, L.E., compilers, 2012, *Geologic Time Scale v. 4.0*: Geological Society of America, doi: 10.1130/2012.CTS004R3C.

Walker, J.D., Geissman, J.W., Bowring, S.A., and Babcock, L.E., compilers, 2018, *Geologic Time Scale v. 5.0*: Geological Society of America, https://doi.org/10.1130/2018.CTS

Web references

Apollo missions:

nasa.gov/mission_pages/apollo/missions/index.html

Jean de Charpentier:

historyofinformation.com/detail.php?id=2384

lindahall.org/jean-de-charpentier/

Jean-Pierre Perraudin:

peoplepill.com/people/jean-pierre-perraudin/

earthobservatory.nasa.gov/features/Paleoclimatology

Louis Agassiz:

geo.mtu.edu/KeweenawGeoheritage/Glaciers/Louis_Agassiz.html

ucmp.berkeley.edu/history/agassiz.html

Pierre Martel:

universetoday.com/tag/earths-climate/

shipseducation.net/glaciers/EarlyIdeas.htm

Red granite:

geology.wisc.edu/homepages/geolib/public_html.old/state.html

University of Wisconsin-Madison: C.K. Leith Library of Geology and Geophysics

Photos

Eagle River Falls flows over the top of the Portage Lake Volcanics. It is located in the Upper Peninsula of Michigan, along M-26. Photo was taken by the author in 2020.

Miller Beach looking out into Lake Michigan. It is located in Gary, Indiana. Photo was taken by the author in 2021.

Dakota Formation along I-25 in New Mexico outside Las Vegas. The author appears in the photo. Photo was taken by Sarah M.H. Baumann in 2020.

A geologically recent lava tube at El Calderon in El Malpais, New Mexico. Photo was taken by the author in 2020.

Along Zuni Canyon Road in New Mexico. Photo was taken by the author in 2020.

Grand Canyon along the South Rim in Arizona. Photo was taken by the author in 2020.

Author's foot is on a slightly older than one billion-year-old stromatolite at Horseshoe Harbor in the Upper Peninsula of Michigan. Photo was taken by the author in 2020.

Author along side a piece of copper bearing conglomerate at the A.E. Seman Mineral Museum in Houghton, Michigan. Photo was taken by Sarah M.H. Baumann in 2021.

Ice on Lake Superior at McLain State Park in the Upper Peninsula of Michigan. Photo taken by the author in 2021.

Sunset Crater is just north of Flagstaff Arizona and is composed of mafic lavas less than 1000 years old! Photo was taken by the author in 2020.

Top of Grinnell Trail in Glacier National Park Montana. Photo was taken by Eamon Section in 2019.

Yellow Mounds Overlook in the Badlands of South Dakota.
Photo was taken by the author in 2019.

Devils Tower Wyoming. The rock is a phonolite porphyry and is igneous in origin. Photo was taken by the author in 2019.

Iceberg Trail at Glacier National Park with the author in the photo. Photo was taken by Sarah M.H. Baumann in 2019.

Climbers on the Baraboo Quartzite at Devils Lake in Wisconsin. Photo was taken by the author in 2019.

Archean rocks that form the cliffs along Agawa Trail in Ontario. Photo was taken by the author in 2019.

Mesoproterozoic iron carbonate called the Gunflint Formation just outside of Thunder Bay, Ontario along Trans-Canada 17. Photo was taken by the author in 2019.

Mesoproterozoic sedimentary rocks of the Sibley Group with a dark igneous sill at Kama Hill along Trans-Canada 17 in Ontario. Photo was taken by the author in 2019.

Gray lava on top of a red sedimentary rock along State Route 61 several miles west-southwest of Grand Marias, Minnesota. Photo was taken by the author in 2019.

Paleoproterozoic tillite of the Gowganda Formation along Provincial Route 129 in Ontario. Photo taken by the author in 2017.

Precambrian meta-igneous rocks along Provincial Route 129 in Ontario. Photo was taken by the author in 2017.

Pennsylvanian sandstone at Turkey Run State Park in Indiana. Photo was taken by the author in 2017.

Volcanics of the Mid-continent Rift at Lake of the Clouds in Porcupine Mountain State Park in the Upper Peninsula of Michigan. Photo taken by the author in 2017.

Large white quartz body at Horserace Rapids in the Upper Peninsula of Michigan. Photo was taken by the author in 2016.

Precambrian till of the Gowganda Formation in Ontario Canada, Notice the large lighter colored boulders (mostly granite) within the darker, fine grained matrix (mostly argillite). $2 coin for scale at the center left. Photo was taken by the author.

Van His rock, located along Route 136, about ¾ of a mile north of Rock Springs Wisconsin. The rock is made of the Precambrian Baraboo Quartzite. Photo was taken by the author.

The Precambrian Jacobsville Sandstone near Assinins Michigan, along the west side of L'Anse Bay in the Upper Peninsula. Photo was taken by the author in 2015.

The unconformity between the Precambrian Jacobsville Sandstone (foreground) and the Precambrian Mesnard Quartzite (background under the trees) at Lake Superior in Marquette Michigan. Photo was taken by the author in 2015.

Split Rock is a granite glacial erratic, with a tree growing out of it. It is located in the Upper Peninsula of Michigan, along U.S. Route 45. Photo was taken by the author in 2015.

Further reading

Books/Textbooks

"The Man Who Found Time" by Jack Repcheck

"A History of Geology" by Gabriel Gohau

"Historical Geology" by Reed Wicander and James S. Monroe

"Annals of the Former World" by John McPhee

Conference Papers

"Toward a History of Geology" 1967 New Hampshire Inter-disciplinary Conference on the History of Geology, edited by Cecil J. Scheer

Epilogue

I wanted this book to be more than just a historical account. As a geologist, I also felt the need to demonstrate that geology didn't just come about one day and that it is a complex subject with many subdisciplines. Over the past few decades there seems to be a growing anti-science push in most of western civilization. I don't know, maybe it has always been there, and the internet has just given a platform to the ignorant. I think it is important that you understand my thought process and why this book contains some of the comments it does.

Before I get all philosophical on you, I do need to make a clarification. Geology isn't just looking at rocks and going "ooooo....ahhhhhh". It's more than just rock identification. Geologists have to take a lot of math and other sciences as well, for a reason. Concepts like structural geology, radiometric dating, and plate tectonics are extremely maths intensive and based in other sciences. I do realize that most people's exposure to geology is introductory rock identification and basic concepts. This leads to the assumption that all concepts in geology can be easily understood. It's one of the reasons why young earth creationists fail. They don't even understand the core principles in geology, although they think they are experts. I encourage self-learning and curiosity. But it is not as easy as many people think it is to self-educate. Especially if you don't know how to vet your sources and distinguish between actual science and someone pushing pseudoscience.

I was hoping to convey the idea that it was a slow process out of assumption based explanations of the world, into the realm of science. Don't get me wrong, science cannot figure out everything. It also does not seek absolute truth or proof. People will often toss those words around in a generic manner, and I guess that's fine, as long as all parties understand. I personally avoid them when talking about science. Science can only come up with explanations based on the evidence at the time. This is why predictive modeling is important. Modeling is a way to verify our methods and concepts. Science won't get you to any absolute truth of the universe, if such a thing even exists. But it will increase our academic and practical knowledge base.

We can only use science to describe natural occurrences based on the testable observations and evidence. Things outside of the observable universe are outside of the reach of science. If you can't test it, you don't worry about it. at least in the present. Many things aren't testable today, but maybe with the advent of new technology they could be. After all, that's how we got the theory of plate tectonics.

I also think science has matured to a point where it is now separate from philosophy. I can't tell you exactly when this happened. Not everything has an exact date or line you can point to.

Science's limitations are also its strengths. It has a natural filter that philosophy lacks. In science, if we can't test it, we either toss it aside or shelve it until we can. Philosophy has no such filter. It is open to entertaining any and every idea, no matter how obviously ridiculous. Philosophy also uses a completely different set of tools than science.

I do not think philosophy is worthless. I do recognize its ability to address concepts science cannot. I personally just don't see the point of talking and arguing indefinitely about a topic where not only no consensus, but no new knowledge can be gained. Philosophy, by its nature is far more static than science, which generally progresses. I know maths and science all had their origins in philosophy, but that no longer matters outside of a historical context. Science has grown up, as where philosophy has pretty much remained unchanged for centuries.

I also do not agree with people who want to call everything and anything, philosophy. If you are going to toss a blanket over everything that happens or exists as philosophy, just because you want to keep it relevant, you have essentially made the concept completely worthless.

Even us scientists are people. We will hold opinions on things, just as everyone else does. We are not robots. We all don't have the same personal beliefs, morals, or talents; just like everyone else. I feel we want the same thing most people do, an understanding of the world in which we live.

www.ingramcontent.com/pod-product-compliance
Lightning Source LLC
Chambersburg PA
CBHW070804220526
45466CB00002B/543